Berichte zu
Tierarzneimitteln 2008

Inhaltsverzeichnis

1 Pharmazeutische Qualität von Tierarzneimitteln

Eine angemessene Qualität ist neben der Wirksamkeit und der Unbedenklichkeit eine auch durch das Arzneimittelgesetz (AMG) festgeschriebene zentrale Anforderung an ein Arzneimittel.

In der Regel wird man einem Arzneimittel nicht ansehen können, ob es von „guter" oder „schlechter" Qualität ist. Nur in Ausnahmefällen (z. B. wenn eine Tablette bröckelt, eine Injektionslösung durch Verkeimung trüb wird oder ein Pulver durch Feuchtigkeitsaufnahme verklumpt ist) kann der Anwender von Mängeln in der Qualität eines Tierarzneimittels sprechen. Aber ohne eingehende Prüfung ist kaum zu beurteilen, ob z. B.

- *ein Arzneimittel überhaupt den deklarierten Wirkstoff enthält,*
- *der Wirkstoffgehalt dem deklarierten Wert entspricht,*
- *das Tierarzneimittel einen zu hohen Gehalt an Zersetzungsprodukten enthält,*
- *der Wirkstoff in einer kaum löslichen Kristallmodifikation vorliegt und keine ausreichende Bioverfügbarkeit gegeben ist.*

Für den Anwender eines Tierarzneimittels stehen zumeist die beobachtete Wirkung und eventuell auftretende Nebenwirkungen im Vordergrund. Hingegen wird die pharmazeutische Qualität von Tierarzneimitteln grundsätzlich vorausgesetzt und kaum angezweifelt.

Die Sicherstellung einer ausreichenden Qualität von Tierarzneimitteln soll einerseits durch den Hersteller und andererseits durch die Zulassungs- und Überwachungsbehörden gewährleistet werden.

Nationale wie auch internationale Rechtsvorschriften legen die dem Stand der Wissenschaft angemessene pharmazeutische Qualität als eine wesentliche Voraussetzung für die Erteilung der Zulassung für ein Tierarzneimittel fest. Das Arzneimittelgesetz (AMG) führt die Sicherstellung der Qualität neben der Wirksamkeit und Unbedenklichkeit als Schwerpunkt in seiner Zweckbestimmung auf. Arzneimittel in mangelhafter Qualität dürfen nach dem AMG weder hergestellt noch in Verkehr gebracht werden.

Das AMG definiert die Qualität als die Beschaffenheit eines Arzneimittels, die nach Identität, Gehalt, Reinheit, sonstigen chemischen, physikalischen, biologischen Eigenschaften oder durch das Herstellungsverfahren bestimmt wird (§ 4 Abs. 15 AMG).

Die Anforderungen an die pharmazeutische Qualität von Arzneimitteln sind in den letzten Jahren deutlich angestiegen – dies betrifft sowohl Human- als auch Tierarzneimittel. Ursache hierfür sind sowohl das insgesamt gesteigerte Qualitätsbewusstsein als auch die deutlich verbesserten technischen und analytischen Möglichkeiten zur Herstellung und Kontrolle von Arzneimitteln. Nicht zuletzt im Zuge der europäischen und der weltweiten Harmonisierung von Zulassungsanforderungen haben sich die Zulassungskriterien erhöht. Grundsätzlich werden in der EU an Tierarzneimittel hinsichtlich der pharmazeutischen Qualität die gleichen Anforderungen wie an Humanarzneimittel gestellt.

1.1. Grundlagen für die Prüfung und Beurteilung der pharmazeutischen Qualität von Tierarzneimitteln

Die bei der Zulassung eines Tierarzneimittels zu stellenden Anforderungen ergeben sich aus einer Vielzahl von Vorschriften. Die wesentliche Grundlage für die Prüfung und Beurteilung der Unterlagen zur pharmazeutischen Qualität stellen die folgenden Vorschriften dar.

1.1.1 Tierarzneimittelprüfrichtlinien

Die grundlegenden Forderungen des Arzneimittelgesetzes an die pharmazeutische Qualität werden insbesondere durch die Tierarzneimittelprüfrichtlinien nach § 26 AMG präzisiert. Sie enthalten Maßstäbe, die an die Zulassungsunterlagen zu stellen sind. Die nationalen Tierarzneimittelprüfrichtlinien verweisen direkt auf EU-Vorgaben (siehe Anhang I der Richtlinie 2001/82/EG zuletzt geändert durch Richtlinie 2009/9/EG).

1.1.2 Arzneibücher

Bei den Arzneibüchern spielt insbesondere das Europäische Arzneibuch eine maßgebliche Rolle. Aber auch das Deutsche Arzneibuch sowie das Homöopathische Arzneibuch sind hier aufzuführen. Die Arzneibücher stellen den Stand der wissenschaftlichen Erkenntnisse dar. Sie sind eine Sammlung anerkannter pharmazeutischer Regeln über die Qualität, Prüfung, Lagerung, Abgabe und Bezeichnung von Arzneimitteln und den bei ihrer Herstellung verwendeten Stoffen. Gemäß § 55 AMG dürfen Arzneimittel nur hergestellt oder in Verkehr gebracht werden, wenn sie den Anforderungen des Arzneibuchs entsprechen.

1.1.3 EU-Leitlinien zur Qualität

Auf EU-Ebene gibt es eine Vielzahl von Leitlinien (Guidelines) zur pharmazeutischen Qualität von Tierarzneimitteln. Diese EU-Leitlinien werden von der Joint CHMP/CVMP Quality Working Party erarbeitet. Diese gemeinsame Arbeitsgruppe des Committee for Human Medicinal Products (CHMP) und des Committee for Veterinary Medicinal Products (CVMP) tagt regelmäßig bei der Europäischen Arzneimittelagentur (EMEA) in London. Die Leitlinien sowie deren Entwürfe sind auf der Homepage der Europäischen Arzneimittelagentur veröffentlicht.

Ein Teil dieser EU-Leitlinien, vor allem solche, die sich an Tierarzneimittel mit neuen Wirkstoffen richten, sind im Rahmen eines internationalen Harmonisierungsprozesses (International Cooperation on Harmonisation of Technical Requirements for Registration of Veterinary Products; VICH) entstanden. Hier haben sich die Wirtschaftblöcke EU, USA und Japan auf einheitliche Zulassungskriterien geeinigt. Im Bereich der pharmazeutischen Qualität von Tierarzneimitteln ist der Harmonisierungsprozess besonders weit fortgeschritten.

1.1.4 Verordnungen und Bekanntmachungen

Insbesondere auf nationaler Ebene sind eine Vielzahl von weiteren arzneimittelrechtlichen Bestimmungen (z. B. Verordnungen wie AMRadV, AMG-TSE-VO, ArzneimittelfarbstoffVO, AflatoxinVerbotsVO) sowie Bekanntmachungen der Zulassungsbehörde zu berücksichtigen.

1.2
Nachweis der pharmazeutischen Qualität in den Zulassungsunterlagen

In seiner Zulassungsdokumentation, die dem BVL als zuständiger Bundesoberbehörde zur Prüfung vorzulegen ist, muss der Antragsteller belegen, dass das beantragte Tierarzneimittel die nach den anerkannten pharmazeutischen Regeln notwendige Qualität aufweist. Gleichzeitig ist nachzuweisen, dass das Arzneimittel hinsichtlich seiner pharmazeutischen Qualität nach dem gesicherten Stand der wissenschaftlichen Erkenntnisse ausreichend geprüft worden ist.

Der Aufbau der Zulassungsdokumentation für Tierarzneimittel entspricht den Vorgaben der EU (sogenanntes „Notice to Applicants-Format (NTA-Format)"). Der für Humanarzneimittel seit einigen Jahren neu vorgeschriebene, international harmonisierte Aufbau im sogenannten „CTD-Format" (Common Technical Document) wird in der Regel auch für die Tierarzneimittelzulassung akzeptiert; dies gilt jedoch nur für die Unterlagen zur pharmazeutischen Qualität.

Teil II der Zulassungsdokumentation eines Tierarzneimittels (NTA-Format) betrifft die pharmazeutische Qualität. Entsprechend den Vorgaben der Tierarzneimittelprüfrichtlinien werden im Zulassungsverfahren Unterlagen zu folgenden Themenkomplexen vorgelegt:

1.2.1 Zusammensetzung

Unter dem Kapitel Zusammensetzung ist die vollständige quantitative und qualitative Zusammensetzung des Tierarzneimittels aufzuführen. Dazu gehören auch Farbstoffe (z. B. für die Kapselhüllen), Konservierungsmittel, Antioxidantien, Aromastoffe und Geschmacksverbesserer, Lösungsvermittler, Stoffe zur Einstellung des pH-Wertes oder Schutzgase (bei Parenteralia/Ampullen). Die Behältnisse sind zu beschreiben. Auch die Zusammensetzung der für die klinische Prüfung verwendeten Chargen ist anzugeben.

Insbesondere ist hier die pharmazeutische Entwicklung des Tierarzneimittels darzulegen. Dazu sind Angaben zur Wahl der Zusammensetzung, der Darreichungsform, des Behältnisses und des Herstellungsverfahrens zu machen und entsprechende Begründungen zu geben.

Ein typisches Beispiel für eine Fragestellung, die unter der pharmazeutischen Entwicklung abgehandelt wird, betrifft den Zusatz von Konservierungsmitteln: Früher wurden Injektionszubereitungen häufig hohe Mengen an Konservierungsmitteln zugesetzt – z. T. auch um Mängel bei der Herstellung zu kompensieren. Von Konservierungsmitteln wie Benzylalkohol oder PHB-Ester können bekanntermaßen Risiken für die Zieltiere ausgehen. Daher ist vom Antragsteller in den Zulassungsunterlagen nachzuweisen, dass der Zusatz eines Konservierungsmittels überhaupt erforderlich ist. Der Konservierungsmittelgehalt darf auch nur so hoch sein, wie unbedingt erforderlich. Er muss jedoch ausreichend sein, um z. B. ein Keimwachstum in der Injektionslösung nach Anbruch eines Mehrdosenbehältnisses zu verhindern.

1.2.2 Herstellung

Der Antragsteller muss hier die Herstellungsformel für die vorgesehene Chargengröße angeben. Eine Beschreibung des Herstellungsverfahrens mit der Angabe der Inprozesskontrollen gehört ebenso in die Zulassungsunterlagen, wie die Ergebnisse von experimentellen Untersuchungen zur Validierung des Herstellungsverfahrens (z. B. Vorlage von Ergebnissen, die belegen, dass das vorgesehene Sterilisationsverfahren geeignet ist, die Sterilität des beantragten Tierarzneimittels sicherzustellen).

1.2.3 Kontrolle der Ausgangsstoffe

In dem Kapitel „Kontrolle der Ausgangsstoffe" müssen alle Bestandteile des Arzneimittels (Wirkstoffe, Hilfsstoffe, Behältnisse und Verschlüsse) beschrieben werden. Wenn diese Ausgangsstoffe im Europäischen oder Deutschen Arzneibuch aufgeführt sind, müssen sie den dort festgelegten Anforderungen entsprechen. Ggf. sind die Arzneibücher anderer EU-Staaten oder die von Drittstaaten heranzuziehen. Gibt es keine Monographien in Arzneibüchern, sind in den Zulassungsunterlagen Spezifikationen und Prüfmethoden für die Ausgangsstoffe ausführlich zu beschreiben.

Wirkstoffe für die Arzneimittelherstellung können in der Regel nie in absolut reiner Form hergestellt werden. Sie enthalten häufig noch Verunreinigungen aus der Synthese (z. B. Syntheseausgangsstoffe, Neben- und Zwischenprodukte, Restlösungsmittel, Katalysatoren).

Das Herstellungsverfahren für einen Wirkstoff kann von Hersteller zu Hersteller stark schwanken (Beispiel: Aminosäuren, die durch völlig unterschiedliche Verfahren gewonnen werden: biotechnologische Verfahren, Hydrolyse von Proteinen oder klassische chemische Synthese). Dementsprechend kann das Verunreinigungsprofil im Wirkstoff stark variieren. Aber auch bei gleichem Syntheseverfahren kann es bei verschiedenen Wirkstoffherstellern zu unterschiedlichen Verunreinigungen kommen. Selbst beim gleichen Wirkstoffhersteller können von Charge zu Charge unterschiedliche Verunreinigungen auftreten. Es ist daher besonders wichtig, dass alle vorhandenen und möglichen Verunreinigungen im Wirkstoff durch entsprechende Analysenverfahren erkannt werden und ihr Gehalt adäquat limitiert wird.

Für neue Wirkstoffe sind in der Regel eine Identifizierung (Strukturaufklärung) oberhalb bestimmter Schwellenwerte sowie eine Qualifizierung (Nachweis der Unbedenklichkeit) oberhalb von 0,5 % erforderlich.

Auch bei Wirkstoffen, die durch eine Monographie eines Arzneibuchs beschrieben sind, ist in den Zulassungsunterlagen nachzuweisen, dass die Methoden der Arzneibuchmonographie geeignet sind, die Qualität des Wirkstoffs zu kontrollieren. Wirkstoffhersteller können die Übereinstimmung ihres Wirkstoffes mit den Anforderungen des Europäischen Arzneibuchs auch durch ein entsprechendes Zertifikat des Europäischen Arzneibuchs (CEP) nachweisen.

Active Substance Master File (ASMF)
Grundsätzlich müssen einem Antragsteller alle wesentlichen Informationen zur Herstellung des in seinem Arzneimittel verwendeten Wirkstoffs bekannt sein (z. B. Syntheseweg, verwendete Katalysatoren, Restlösungsmittel). Nur so kann er in vollem Umfang die erforderliche Verantwortung für sein Arzneimittel übernehmen.

Bei Wirkstoffhersteller und Antragsteller für ein zuzulassendes Arzneimittel handelt es sich häufig um verschiedene, getrennt voneinander agierende Firmen. Für den Wirkstoffhersteller gehören Details zur Synthese seines Wirkstoffs zumeist zu den besonders schützenswerten Firmengeheimnissen. Diese sollen dem Antragsteller bzw. Inhaber der Zulassung oder anderen an der Herstellung des Arzneimittels beteiligten Firmen aus nahe liegenden Gründen nicht bekannt werden.

Diesem Umstand trägt das ASMF-Verfahren Rechnung. Ein ASMF besteht aus einem offenen Teil (open part oder applicants part) und einem vertraulichen Teil (closed part oder restricted part), der Details der Synthese enthält. Der offene Teil des ASMF wird im Rahmen der Zulassungsdokumentation vom Antrag-

steller für das Tierarzneimittel bei der Zulassungsbehörde eingereicht. Der vollständige ASMF mit dem vertraulichen Teil wird vom Wirkstoffhersteller direkt bei der Zulassungsbehörde vorgelegt.

1.2.4 Kontrolle der Zwischenprodukte

Wenn bei der Herstellung eines Arzneimittels Zwischenprodukte entstehen (z. B. Drageekerne, die anschließend noch mit einem Überzug versehen werden) sind die Prüfungen in diesem Kapitel der Zulassungsunterlagen zu beschreiben.

1.2.5 Kontrolle des Fertigproduktes

Für das Fertigprodukt wird vom Antragsteller eine Freigabespezifikation vorgelegt. Diese Spezifikation enthält alle wesentlichen Merkmale des Arzneimittels; bei quantifizierbaren Merkmalen sind geeignete Grenzwerte festzulegen.

Bei einer Injektionslösung können das z. B. folgende Merkmale sein: Aussehen, Abwesenheit sichtbarer Partikel, Farbe, Klarheit, entnehmbares Volumen, Dichte, pH-Wert, Identität des Wirkstoffs und des Konservierungsmittels, Gehalt des Wirkstoffs und des Konservierungsmittels, Gehalt an Verunreinigungen und Abbauprodukten und Sterilität.

In der Regel werden alle Merkmale der Freigabespezifikation vom Hersteller eines Tierarzneimittels bei der Freigabe jeder einzelnen Charge untersucht und in Analysenzertifikaten dokumentiert. Die bei der Freigabeuntersuchung verwendeten Prüfmethoden sind in den Zulassungsunterlagen ausführlich zu beschreiben und zu validieren.

1.2.6 Haltbarkeit

Ein zentrales Element der Zulassung ist die Prüfung und Festlegung der Haltbarkeit von Wirkstoff und Fertigarzneimittel. Für das Fertigarzneimittel hat ein Antragsteller eine Haltbarkeitsspezifikation vorzulegen, die nur in begründeten Fällen von der Freigabespezifikation abweichen darf. Durch Vorlage entsprechender Untersuchungsergebnisse hat der Antragsteller nachzuweisen, dass diese Spezifikationen über die vorgesehene Dauer der Haltbarkeit eingehalten werden. Während früher die Haltbarkeitsuntersuchungen unter sehr unterschiedlichen und z. T. nicht standardisierten Bedingungen durchgeführt worden sind, werden heute Klimakammern mit definierten Bedingungen eingesetzt.

Die Haltbarkeitsuntersuchungen sind unter international harmonisierten Bedingungen (v. a. Lagerungstemperatur, relative Feuchte (RH)) durchzuführen. In der Regel sind das Langzeituntersuchungen bei 25 °C ± 2 °C / 60 % RH ± 5 % RH oder bei 30 °C ± 2 °C / 65 % RH ± 5 % RH sowie beschleunigte Untersuchungen über mindestens 6 Monate bei 40 °C ± 2 °C / 75 % RH ± 5 % RH. Die so eingelagerten Arzneimittelmuster werden in bestimmten Abständen (anfangs vierteljährlich, später halbjährlich, danach in jährlichen Abständen) auf alle Merkmale der Spezifikation geprüft. Basierend auf den Ergebnissen der

Haltbarkeitsuntersuchungen werden die Dauer der Haltbarkeit und die Lagerungshinweise für das Tierarzneimittel festgesetzt.

Sofern zutreffend, sind auch die Haltbarkeit nach Anbruch und die Haltbarkeit nach Rekonstitution (z. B. Herstellung der gebrauchsfertigen Lösung) im Rahmen der Zulassungsunterlagen nachzuweisen.

1.2.7   *Sonstige Informationen*

In diesem Teil der Zulassungsdokumentation sind Informationen enthalten, die in den vorherigen Teilen nicht erfasst sind. Dies betrifft z. B. spezielle Angaben zu Arzneimittel-Vormischungen, die zur Herstellung von Fütterungsarzneimitteln dienen (z. B. Einmischraten, Eignung von Mischfuttermitteln, Haltbarkeit des Fütterungsarzneimittels).

1.3
Überwachung von Herstellern

Die Überwachung von Tierarzneimittelherstellern obliegt in Deutschland den zuständigen Landesbehörden. Diese kontrollieren in regelmäßigen Intervallen, ob die Herstellung ordnungsgemäß verläuft und die im Verkehr befindlichen Arzneimittel die erforderliche Qualität besitzen (z. B. Überprüfung durch Arzneimitteluntersuchungsstellen der Länder). Voraussetzung für die Herstellung von Tierarzneimitteln ist eine Herstellungserlaubnis der zuständigen Landesbehörde. Diese ist an das Vorhandensein geeigneter Räume und Einrichtungen für die Herstellung, Prüfung und Lagerung von Arzneimitteln gebunden. Entsprechend qualifiziertes und verantwortliches Personal muss ebenfalls vorhanden sein.

2 Regulatory Aspects Concerning the Efficacy and Safety Assessment of Veterinary Homeopathic Products and Herbal Medicines[1]

Andrea Golombiewski

Federal Office of Consumer Protection and Food Safety, Berlin

Correspondence to: Dr. A. Golombiewski, BVL, Ref. 303, Mauerstraße 39–42, D-10117 Berlin, Germany, Tel. 030 18 444 30317, e-mail: andrea.golomiewski@bvl.bund.de

Pet owners, farmers and veterinarians are becoming increasingly interested in complementary and alternative medicine, including homeopathy, anthroposophy and use of herbal medicine, primarily due to its reputation of being a "gentle" treatment option.

Regulatory issues for all veterinary medicinal products are described in Directive 2001/82/EC on the Community code relating to veterinary medicinal products, amended by Directive 2004/28 EC, which forms the basis for the national drug laws in the member states. As for any veterinary medicinal product there is the option of the presentation of a full dossier for the application for marketing authorisation for homeopathic/anthroposophic or herbal products, including a full set of toxicological, target animal safety and efficacy data, or to demonstrate that the product has been in well-established veterinary use within the Community for at least ten years, with recognised efficacy and an acceptable level of safety.

The above Directive offers particular provisions for veterinary homeopathics which enables the establishment of a simplified procedure for registration in the member states. Following this procedure, indications must not be named but safety has to be guaranteed. Difficulties in assessment arise from the principles of treatment or the nature of the drug, depending on the type of the alternative product.

For homeopathic or anthroposophic products a classical demonstration of efficacy, including dose justification and field trials, is extremely difficult due to the specific rules of these highly individualized types of treatment. Therefore, the simplified procedure is the usual way to apply for a marketing authorisation. Safety, however, can be an issue depending on the substances contained, their dilution and the administration route.

Herbal products are complex mixtures with an often variable or even partly unknown composition, while toxicology, safety and efficacy studies require standardised products. However, for some well-known plant derived drugs, there is information available based on the experience of use. An assessment of these products has to take into account all available information, including monographs developed for use in humans, to make a reliable conclusion regarding efficacy and safety in the target species.

[1] In: Abstracts of the 11th European Association for Veterinary Pharmacology and Toxicology, Leipzig, Germany, 12–16 July 2009. Journal of Veterinary Pharmacology and Therapeutics, Volume 32 Issue s1, Pages 1–270 (August 2009) Special Issue

Bundesamt für
Verbraucherschutz und
Lebensmittelsicherheit

Structure of the presentation

- **Introduction**

- **Legal basis**

- **Assessment of veterinary homoeopathic products**

- **Assessment of veterinary herbal products**

- **Conclusions**

Andrea Golombiewski • Seite 2

Bundesamt für
Verbraucherschutz und
Lebensmittelsicherheit

Legal basis - general

**Regulatory requirements for marketing authorization of veterinary medicinal products (VMP):
Directive 2001/82 EC as amended**

What are the options in general?

Full dossier	Well-established vet use
Toxicological data	Acceptable level of safety
Target animal safety data	
Efficacy data	Recognised efficacy

Andrea Golombiewski • Seite 3

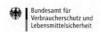

Legal basis - homoeopathics

What are the options
for homoeopathic / anthroposophic VMPs?

Full dossier	Well-established vet use	Simplified procedure
Toxicological data	Acceptable level of safety	Product must not pose a risk for the animal
TAS data		
Efficacy data	Recognised Efficacy	No efficacy data

Andrea Golombiewski • Seite 4

Assessment of homoeopathic VMPs (1)

Efficacy assessment-
Why can it be omitted for homoeopathics?

Efficacy requires proof for
specific indications / specific dosing regimen.

Homoeopathy is a highly individualized treatment.

Double-blind randomized field trials are difficult
- if not impossible for this type of product.

Indications must not be listed in the product literature, but the
homoeopathic community should know about the fields of use.

Andrea Golombiewski • Seite 5

Bundesamt für
Verbraucherschutz und
Lebensmittelsicherheit

Assessment of homoeopathic VMPs (2)

Target animal safety assessment-
Why is safety an issue in homoeopathics?

Target animal safety has to be guaranteed to come to a positive conclusion in the risk-benefit-evaluation.

- „Homoeopathic" does not automatically mean „highly diluted"
- „Homoeopathic" does not automatically mean „no harm"
- Allergic reactions
- Oral intake of external preparations (by licking)
- Local tolerance of injectables: injection volume, preservatives
- Owner's overreliance on both efficacy and „gentleness" of treatment („Over-the-counter"!)

Andrea Golombiewski • Seite 6

Bundesamt für
Verbraucherschutz und
Lebensmittelsicherheit

Assessment of homoeopathic VMPs (3)

Target animal safety assessment-
How is this performed / addressed in homoeopathics?

Toxicological aspects:
Expert report
Information available in the public domain

Wording of product information:
In Germany, information for the user includes
- which symptoms require to see a vet
- to ask a vet to provide training to user before giving injections
- to ask a vet before application to a pregnant /lactating animal or before combination of products

Andrea Golombiewski • Seite 7

Bundesamt für
Verbraucherschutz und
Lebensmittelsicherheit

Legal basis – herbal products

What are the options for herbal VMPs?

Full dossier	Well-established vet use	In Germany*: Traditional herbal VMPs
Toxicological data	Acceptable level of safety	Must be safe when used as claimed
TAS data		
Efficacy data	Recognised Efficacy	Efficacy based on experience of use and plausibility

*Traditional herbal products are not mentioned in Directive 2001/82 as amended,
 but in the counterpart for humans (Directive 2001/83 as amended) and in German drug law.

Andrea Golombiewski • Seite 8

Bundesamt für
Verbraucherschutz und
Lebensmittelsicherheit

Assessment of herbal VMPs (1)

**Efficacy assessment-
How is this performed for (traditional) herbal VMPs?**

In a „regular" application (full dossier, well-established use), efficacy
   has to be proven for the claimed indication at the claimed dose.
But:
Herbals are complex mixtures with variable/unknown composition.
Efficacy studies require standardised products.

In an application for a traditional herbal VMP, efficacy is based on
   experience of use and plausibility.
   - 30 years of veterinary use, 15 of which in the EU
   - plausibility = pharmacological properties or experiences of use

Andrea Golombiewski • Seite 9

Assessment of herbal VMPs (2)

Target animal safety assessment-
Why is safety an issue in (traditional) herbal VMPs?

- Target animal safety has to be guaranteed to come to a positive conclusion in the risk-benefit-evaluation.
- „Herbal" does not automatically mean „no harm"

What are the specific challenges?

- Complex mixtures:
  Toxicological tests not designed for such mixtures
- Not always standardized:
  Target animal safety studies require standardized products
- Inter-species extrapolation?

Andrea Golombiewski • Seite 10

Assessment of herbal VMPs (3)

Target animal safety assessment - How is this performed / addressed in (traditional) herbal VMPs?

In a „regular" application (full dossier, well-established use), safety <u>has to be proven.</u> = full set of studies needed.

In an application for a traditional herbal VMP, safety is based on
- Experiences of use
- Expert report
- Information available in the public domain

Andrea Golombiewski • Seite 11

Bundesamt für
Verbraucherschutz und
Lebensmittelsicherheit

Conclusion

Alternative therapies play a role in veterinary medicine

Assessment using the usual rules of marketing authorisation procedure difficult / impossible

Special simplified procedures may be used for veterinary homoeopathic or (in Germany) traditional herbal products

Lack of information does not mean absence of safety concerns

Andrea Golombiewski • Seite 12

3 Clinical Studies Testing Veterinary Pharmaceuticals: The Regulatory Perspective

Gesine Hahn

Federal Office of Consumer Protection and Food Safety, Berlin

Correspondence to: Dr. G. Hahn, BVL, Ref. 303, Mauerstraße 39–42, D-10117 Berlin, Germany, Tel. 030 18 444 30300, e-mail: gesine.hahn@bvl.bund.de

Definition of clinical studies

„The purposes of clinical trials are to demonstrate or substantiate the effect of a vmp after administration of the recommended dosage to specify ist indications and contraindications according to species, age, breed and sex. Its directions for use, any side effects which it may have and ist safety and tolerance under normal conditions of use"

(Annex I of Directive 2001/82/EC)

Gesine Hahn, 21.06.2007

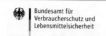
Bundesamt für
Verbraucherschutz und
Lebensmittelsicherheit

Purpose of Clinical Studies

- Confirmatory (e.g. efficacy and safety under practical use conditions)

- Exploratory (e.g. dose finding, target animal safety)

- Composite, i.e. further exploratory analyses to explain the study results and to suggest further hypotheses for research

Gesine Hahn, 21.06.2007

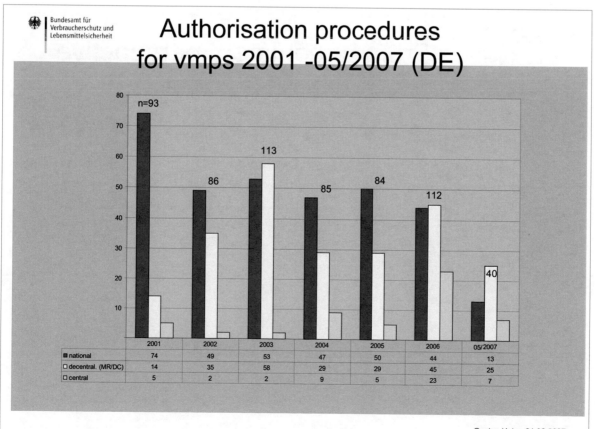

	2001	2002	2003	2004	2005	2006	05/2007
■ national	74	49	53	47	50	44	13
□ decentral. (MR/DC)	14	35	58	29	29	45	25
□ central	5	2	2	9	5	23	7

Gesine Hahn, 21.06.2007

Bundesamt für
Verbraucherschutz und
Lebensmittelsicherheit

VICH GL 9

- Scientific quality standard for organizing, carrying out and reporting of clinical studies

- Standard terminology

- Agreed between pharmaceutical industry and regulators

- Global approval in the European Union (EU), Japan and the U.S. facilitates the mutual acceptance of clinical data by the relevant regulatory authorities.

Gesine Hahn, 21.06.2007

 Bundesamt für
Verbraucherschutz und
Lebensmittelsicherheit

From a regulatory perspective

- Assessors do not inspect the **conduct** of clinical studies

- Assessment is based on the **documentation** only

Gesine Hahn, 21.06.2007

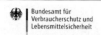
Bundesamt für
Verbraucherschutz und
Lebensmittelsicherheit

Positive aspects for regulators

In principle

- High quality standard
- Transparency
- Confidence
- Facilitation of work
- Control

Gesine Hahn, 21.06.2007

Bundesamt für
Verbraucherschutz und
Lebensmittelsicherheit

Study documentation

- Quality statement (GCP compliance; audit reports or certificates)
- Final study report, safety report, any other reports
- Study protocol
- Animal data listing
- Individual animal records
- Animal owners' informed consent

Gesine Hahn, 21.06.2007

Study protocol (1)

- Clear and precise description of the objective of the study
- Clear description of the study design (control methods, method of randomization, blinding techniques)
- Precise description of the study population
  - inclusion/ exclusion criteria
  - Concomitant treatments (permitted, not permitted)
  - Animal husbandery and feeding management
- Treatments, clear identification of the IVPP and control product)

Gesine Hahn, 21.06.2007

Study protocol (2)

- Efficacy assessment
  - Justification of primary/ secondary endpoints
  - clinical observations and other special tests or analyses
  - Timing and frequency
  - Post-observation period
  - Scoring systems and measurements need to be clearly defined
  - Methods of computing and calculating effects
- Monitoring of adverse events
  - Sufficient frequency of observations
  - Any corrective action

Gesine Hahn, 21.06.2007

Bundesamt für
Verbraucherschutz und
Lebensmittelsicherheit

Study protocol (3)

- Statistics
  - Clear description of the statistical methods including hypothesis to be tested, the parameters to be estimated, the assumptions to be made and the level of significance, the experimental unit and the statistical models
  - The planned sample size should be justified in terms of the target population, the power of the study and pertinent clinical considerations
  - Handling of missing data and outliers

Note: The purpose of the study should be to get a statistically significant <u>and</u> clinically relevant answer

Gesine Hahn, 21.06.2007

Bundesamt für
Verbraucherschutz und
Lebensmittelsicherheit

Study protocol (4)

- Supplements
  - Any change of the protocol to be clearly described and justified (protocol amendment)
  - Any deviation from the protocol to be reported and whether or not this has an impact of the results of the study
  - SOPs that are specific to the study

Gesine Hahn, 21.06.2007

Bundesamt für
Verbraucherschutz und
Lebensmittelsicherheit

Final Study Report

- Format as outlined in part 7 of VICH GL 9
- Accurate and comprehensive description of the study
- Compliance with the protocol
- Presentation <u>and</u> critical evaluation of the study results
- Reporting of adverse events (causality, severity and frequency)

Gesine Hahn, 21.06.2007

Bundesamt für
Verbraucherschutz und
Lebensmittelsicherheit

Problems seen in assessing dossiers

Although GCP standards established since 1992 in the EU, the quality of pivotal clinical studies included in MA dossiers is still variable …

Gesine Hahn, 21.06.2007

Bundesamt für
Verbraucherschutz und
Lebensmittelsicherhe

Problems seen in assessing dossiers

- Non GCP studies are still submitted in dossiers although guidance on GCP is available in the EU since 1992 and VICH GCP since 2001
- Brief GCP compliance statement, but study obviously not performed according to GCP

>> Supportive information

Gesine Hahn, 21.06.2007

Bundesamt für
Verbraucherschutz und
Lebensmittelsicherhei

Problems seen in assessing dossiers

- GCP study, but incomplete/ inadequate documentation, e.g.
    - missing study protocol incl. relevant appendices, individual animal data listing,
    - Study protocol and/or FSR not clearly or differently structured,
    - Deviating/ contradictory information in the protocol, final study report or data listing.

Gesine Hahn, 21.06.2007

Bundesamt für
Verbraucherschutz und
Lebensmittelsicherheit

Problems seen in assessing dossiers

– Statistics:

Description and justification of statistical methods in the study protocol often not sufficient

Change of or additional statistical methods when analysing the results

Change in primary parameter after study termination

>> Many questions

Gesine Hahn, 21.06.2007

Bundesamt für
Verbraucherschutz und
Lebensmittelsicherheit

Conclusion

- GCP is an important quality standard intended to ensure accuracy, integrity and correctness of data, in principle

- Other relevant guidelines should be considered as well, depending on the product, claim etc

- It is in the responsibility of the Applicant to follow the guidelines and to consult regulatory authorities in any case of uncertainty *a priori*

Gesine Hahn, 21.06.2007

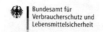
Bundesamt für
Verbraucherschutz und
Lebensmittelsicherheit

Decision on approval

- Based on overall risk-benefit evaluation

- Pivotal clinical studies should be performed according to the state of the art which includes GCP principles

Gesine Hahn, 21.06.2007

4 Pharmacovigilance Inspections: The German Experience

Cornelia Ibrahim

Federal Office of Consumer Protection and Food Safety (BVL), Berlin

Correspondence to: Dr. C. Ibrahim, BVL, Ref. 304, Mauerstraße 39–42, D-10117 Berlin, Germany, Tel. 030 18 444 30400, e-mail: cornelia.ibrahim@bvl.bund.de

Bundesamt für Verbraucherschutz und Lebensmittelsicherheit

Directive 2004/28/EC- main objectives for pharmacovigilance

Pharmacovigilance inspections

⇒ **adequate pharmacovigilance system description requested already at authorisation**

⇒ **pharmacovigilance system inspections**

⇒ **product specific inspections**

⇒ **enforcement of requirements, sanctions**

Dr. Cornelia Ibrahim, Page 2

pharmacovigilance inspections

Guideline on Monitoring of Compliance with Pharmacovigilance Regulatory Obligations and Pharmacovigilance Inspections

- based on: Regulation EC 726/2004 and Directive 2004/28
- public consultation of joint guideline: spring 2006
- now 2 versions: Guideline for Human Medicines published in Volume 9A in January 2007
- Veterinary Guideline adopted by CVMP in January 2007
- published by EU-Commission in April 2007
  will be part of Volume 9B of the „Rules Governing Medicinal Products in the EU" comprising all veterinary pharmacovigilance guidelines, adopted by CVMP in September 2008 for consultation

Dr. Cornelia Ibrahim, Page 3

pharmacovigilance inspections

<u>**Guideline sets out frame for**</u>

- implementation of monitoring pharmacovigilance obligations and inspections
- information to be supplied in marketing authorisation application, giving description of pharmacovigilance system, qualified person and means of notification
- relates to <u>all</u> types of products

Dr. Cornelia Ibrahim, Page 4

Bundesamt für
Verbraucherschutz und
Lebensmittelsicherheit

pharmacovigilance inspections

„Guideline on Monitoring of Compliance with
Pharmacovigilance Regulatory Obligations and
Pharmacovigilance Inspections"

Types of inspections
- routine inspections, 4 years after first marketing
- targeted inspections, when trigger is met
  - not related to specific safety concerns
  - related to specific safety concerns or non-
    compliance

Dr. Cornelia Ibrahim, Page 5

Bundesamt für
Verbraucherschutz und
Lebensmittelsicherheit

pharmacovigilance inspections

„Guideline on Monitoring of Compliance with
Pharmacovigilance Regulatory Obligations and
Pharmacovigilance Inspections"

Types of inspections continued

Pharmacovigilance system inspections
- designed to review system, personnel, facilities
- products may be used as examples to test system
- may be routine or targeted

Dr. Cornelia Ibrahim, Page 6

Bundesamt für
Verbraucherschutz und
Lebensmittelsicherheit

pharmacovigilance inspections

„Guideline on Monitoring of Compliance with
Pharmacovigilance Regulatory Obligations and
Pharmacovigilance Inspections"

Types of inspections continued

Product specific inspections
- focus specifically on a given product
- are usually targeted as a result of identified triggers

Dr. Cornelia Ibrahim, Page 7

Bundesamt für
Verbraucherschutz und
Lebensmittelsicherheit

pharmacovigilance inspections

Experience with inspections in Germany

Preparation before inspection

- written announcement to company with proposal of dates for inspection, usually 2-3 months in advance
- mutual agreement on date
- Invitation of competent federal state authority
- company sends detailed documents on pharmacovigilance system description including relevant SOPs
- preparation of assessors and company
- normally 1 day inspection foreseen

Dr. Cornelia Ibrahim, Page 8

pharmacovigilance inspections

<u>**Experience with inspections in Germany**</u>
<u>**Day of inspection: start with opening remarks and discussion on:**</u>

- structure of company
- national/international sites and relations
- product-portfolio, market position, substance classes, Top 10
- number of substances in phase I-III of authorisation procedure
- number of serious/non-serious reports
- time check and control measures („late cases")
- internal training of staff

Dr. Cornelia Ibrahim, Page 9

pharmacovigilance inspections

<u>**Experience with inspections in Germany**</u>

<u>**Structure of the Pharmacovigilance System**</u>

- qualified person for pharmacovigilance (QPPV) and replacement
- structure of pharmacovigilance unit, number of staff, qualification
- procedure of processing case reports
- description of routes of reporting (call-center, product manager)
- literature research, which database
- number of reports/year (pharmaceuticals/immunologicals)

Dr. Cornelia Ibrahim, Page 10

pharmacovigilance inspections

Experience with inspections in Germany

Quality management system (global/national)
overview, Standard Operating Procedures (SOPs) for
pharmacovigilance

- contracts and communication between different sites and
  organisations in company
- contracts and responsibilities sharing with distributors
- internal/external audits
- follow up of findings, control of deficiencies
- update of SOPs, how often?

Dr. Cornelia Ibrahim, Page 11

pharmacovigilance inspections

Experience with inspections in Germany

Electronic database
- demonstration of system, back-up, emergency procedures in
  case of break-down
- data entry, check ups,
- causality assessment, VEDDRA-coding
- translation to English for Eudravigilance transfer
- correspondence with reporters and authorities
- quality control check, surveillance of time points
- replacement and sharing of responsibilities
- time control PSURs
- archiving procedures of original documents

Dr. Cornelia Ibrahim, Page 12

pharmacovigilance inspections

Experience with inspections in Germany

Crisis management in case of
- **recall**
- **quality defect**
- **urgent safety restriction**
- **information of public on safety issue**

end of inspection
final discussion: summary of findings and first apprehension by authority

Dr. Cornelia Ibrahim, Page 13

pharmacovigilance inspections

Experience with inspections in Germany

after inspection
written assment report by authority assessors with
findings and critical points send to all participants of inspection (company and competent federal state authority) for comments

Final report circulated to company and competent federal state authority
- critical points must be amended within given timeframe
- On case by case basis: re-inspection

Dr. Cornelia Ibrahim, Page 14

pharmacovigilance inspections

Experience with inspections in Germany

No of inspected MAHs:

2006: 2

2007: 7

2008: 10 (6 more planned this year)

Size of inspected MAHs (based on No of national authorisations and/or international parentcompany)

Small: 8

Medium: 5

Large: 6

Dr. Cornelia Ibrahim, Page 15

pharmacovigilance inspections

Experience with inspections in Germany

Findings non-related to the size of inspected MAHs:

• **no 24 h availability of the QPPV**

• **missing SOPs**

• **missing or inappropriate training of staff concerned with pharmacovigilance**

Dr. Cornelia Ibrahim, Page 16

pharmacovigilance inspections

Experience with inspections in Germany

Findings non-related to the size of inspected MAHs:

• missing or incomplete contractural arrangements with other persons or organisations involved in the fulfilment of pharmacovigilance obligations
• missing SOPs

Dr. Cornelia Ibrahim, Page 17

pharmacovigilance inspections

Experience with inspections in Germany

Findings related to the size of inspected MAHs:
Small

• difficulties in calculation PSUR-dates resulting in non or delayed submission of PSURs
• insufficient literature research prior PSURs preparation
• Back up of QPPV often missing or unclear
• definition of SAR unknown
• ABON system unknown

Dr. Cornelia Ibrahim, Page 18

Bundesamt für
Verbraucherschutz und
Lebensmittelsicherheit

pharmacovigilance inspections

Experience with inspections in Germany

Findings related to the size of inspected MAHs:
medium

• **no access to PV-Database of parent-company**
• **complicated processing of SARs between MAH and parent-company resulting in non or delayed submission of SARs**

Dr. Cornelia Ibrahim, Page 19

Bundesamt für
Verbraucherschutz und
Lebensmittelsicherheit

pharmacovigilance inspections

Experience with inspections in Germany

Findings related to the size of inspected MAHs:
Large

• **unskilled employees for research in PV-Database used**
• **non or delayed submission of Third Country reports**
• **insufficient documentation and archiving of original PhV documents (§15 PharmBetrV: it has to be ensured that data are available during the storage period and can be made readable during an adequate time limit)**

Dr. Cornelia Ibrahim, Page 20

Bundesamt für
Verbraucherschutz und
Lebensmittelsicherheit

Future perspectives and harmonisation on EU level

- **Establishment of a separate ad hoc Pharmacovigilance Inspections group at EMEA**
- **important to find harmonised approach between Member States**
- **share information and avoid double work**
- **set up a common database in EU on inspections timing and findings**
- **use inspections as a tool to give advice on good pharmacovigilance practice**
- **enhance compliance with the overall pharmacovigilance system**

Dr. Cornelia Ibrahim, Page 21

5 Elektronische Übermittlung von UAW an das BVL gemäß AMG-AV

Anke Finnah

Bundesamt für Verbraucherschutz und Lebensmittelsicherheit (BVL), Berlin

Korrespondenz an: Dr. A. Finnah, BVL, Ref. 304, Mauerstraße 39–42, D-10117 Berlin, Germany, Tel. 030 18 444 30411, e-mail: anke.finnah@bvl.bund.de

Bundesamt für
Verbraucherschutz und
Lebensmittelsicherheit

Rechtlicher Hintergrund

Verordnung über die elektronische Anzeige von Nebenwirkungen bei Arzneimitteln (AMG-Anzeigeverordnung – AMG-AV)

vom 12. September 2005, in Kraft seit 30. Oktober 2005

§ 2 Abs. 1 AMG-AV:
pharmazeutische Unternehmer sind verpflichtet, Einzelfallberichte gemäß den international geltenden technischen Standards elektronisch gegenüber der zuständigen Bundesoberbehörde (§77 AMG → BVL) anzuzeigen,

Zeitpunkt des Beginns wird von der Bundesoberbehörde bekannt gegeben

Dr. Anke Finnah
04. April 2008   Seite 2

Bundesamt für
Verbraucherschutz und
Lebensmittelsicherheit

Bekanntmachung zur AMG-Anzeigeverordnung – AMG-AV vom 8. Februar 2008

**Bekanntmachung
zur Verordnung
über die elektronische Anzeige
von Nebenwirkungen bei Arzneimitteln
(AMG-Anzeigeverordnung – AMG-AV)**

Vom 8. Februar 2008

Anzeigen von Verdachtsfällen schwerwiegender Nebenwirkungen von Arzneimitteln, die zur Anwendung bei Tieren bestimmt sind und die in den Zuständigkeitsbereich des Bundesamtes für Verbraucherschutz und Lebensmittelsicherheit fallen, sind gemäß AMG-AV vom 12. September 2005 (BGBl. I S. 2775) ab dem 15. März 2008 dem Bundesamt für Verbraucherschutz und Lebensmittelsicherheit ausschließlich in elektronischer Form zu übermitteln. Eine zusätzliche Anzeige in Papierform entfällt (vgl. § 2 Abs. 2 AMG-AV).

Da die Daten zur Übermittlung der „Guideline on Data Elements for the Electronic Submission of Adverse Reaction Reports Related to Veterinary Medicinal Products Authorized in the European Economic Area (EEA) Including Message and Transmission Specifications" (EMEA/CVMP/280/04) entsprechen müssen, stehen folgende Möglichkeiten der elektronischen Meldung zur Verfügung:

1. Dateneingabe in das nationale Online-Meldeformular für pharmazeutische Unternehmer unter www.vet-uaw.de.
2. Dateneingabe in das europäische Datenbanksystem (Eudra Vigilance Veterinary/EVWEB), einschließlich elektronischer Übermittlung in die Inbox der zuständigen Bundesoberbehörde.
3. Dateneingabe in das Online-Formular der Europäischen Arzneimittel-Agentur (MAH Simple Electronic Reporting Form).
4. Datenübermittlung aus firmeneigener Datenbank per Gateway an die Datenbank der zuständigen Bundesoberbehörde.

Weitergehende Informationen zu den verschiedenen Möglichkeiten der elektronischen Anzeige von Verdachtsfällen schwerwiegender Nebenwirkungen von Arzneimitteln, die zur Anwendung bei Tieren bestimmt sind, können unter www.bvl.bund.de eingesehen werden.

Berlin, den 8. Februar 2008
5310 - 04 - 270915

Bundesamt für Verbraucherschutz
und Lebensmittelsicherheit

Im Auftrag
Prof. Dr. R. Kroker

- Verpflichtung zur elektronischen Meldung für alle PU verbindlich seit dem **15. März 2008**

- eine zusätzliche Anzeige in Papierform entfällt

- übermittelte Daten müssen den Vorgaben der „**Guideline on Data Elements** for the Electronic Submission of Adverse Reaction Reports Related to veterinary Medicinal products Authorized in the European Economic Area (EEA) Including Message and Transmission Specifications" (EMEA/CVMP/280/04) entsprechen → XML-Dateien

- 4 verschiedene Möglichkeiten diese Daten elektronisch an das BVL zu übertragen

Dr. Anke Finnah
04. April 2008   Seite 3

Bundesamt für
Verbraucherschutz und
Lebensmittelsicherheit

weitere Informationen unter www.bvl.bund.de

Ausführliche Informationen mit Links zu relevanten Seiten und Dokumenten bei der EMEA sind auf der Homepage des BVL zu finden.

Startseite → Tierarzneimittel → Überwachung und Betreuung

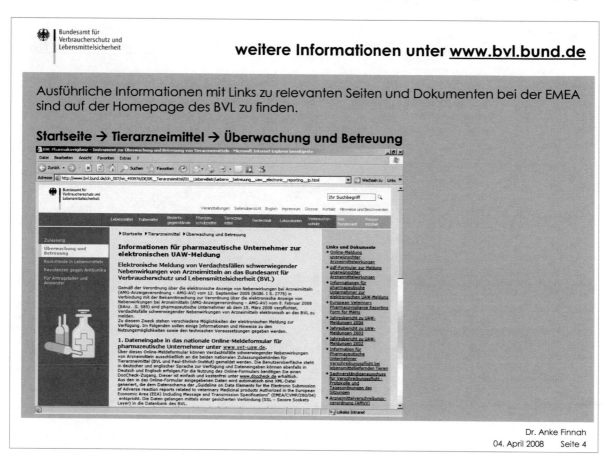

Dr. Anke Finnah
04. April 2008   Seite 4

Bundesamt für
Verbraucherschutz und
Lebensmittelsicherheit

Elektronische Meldung von UAW (Übersicht)

Folgende Möglichkeiten der elektronischen Übertragung von
UAW an das BVL stehen zur Verfügung:

1. UAW-Online-Formular www.vet-uaw.de	eigenes Datenbanksystem nicht notwendig
2. MAH Simple Electronic Reporting Form der EMEA 3. Web-Trader Funktion von EudraVigilance (EVWEB)	eigenes Datenbanksystem nicht notwendig, in Datenbank generierte XML-Dateien können aber verwendet werden
4. ESTRI (Electronic Transfer of Regulatory Information)- Gateway	eigenes Datenbanksystem muss vorhanden sein

Dr. Anke Finnah
04. April 2008   Seite 5

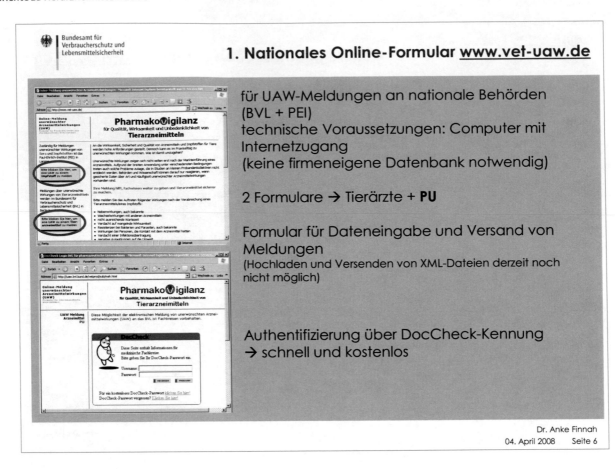

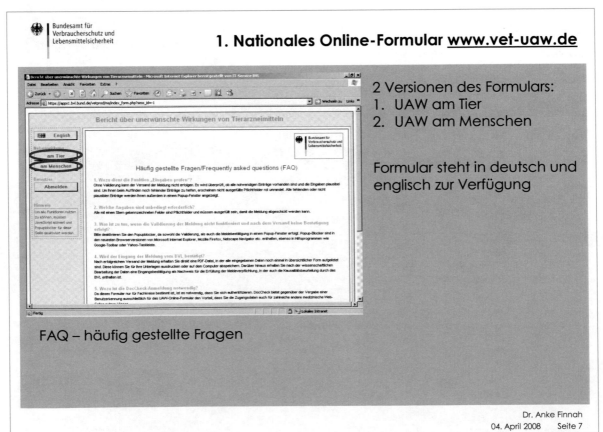

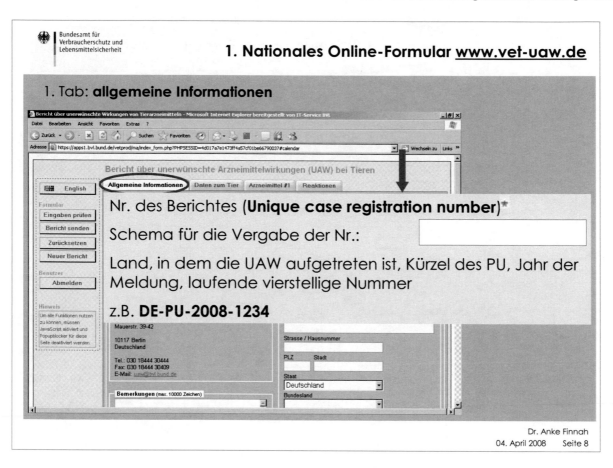

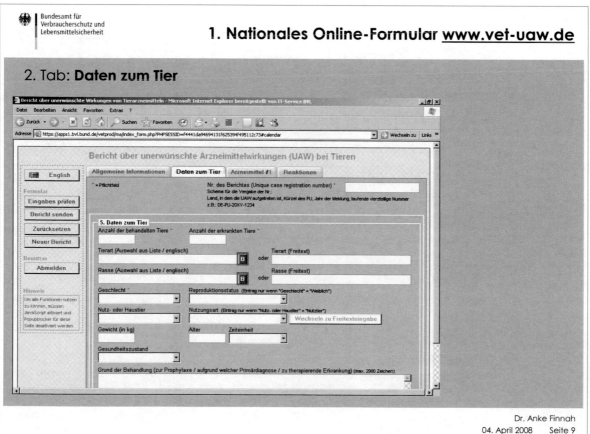

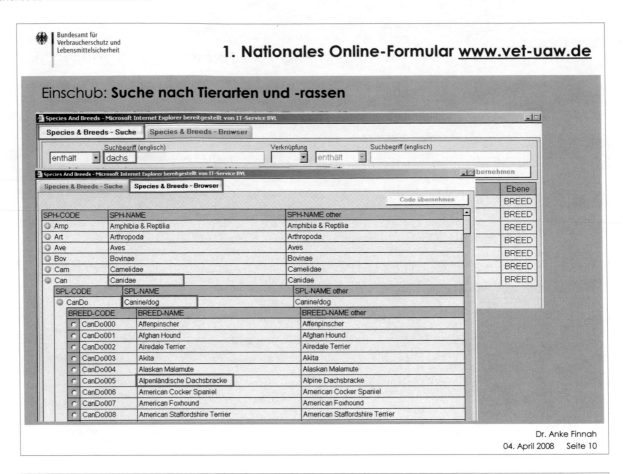

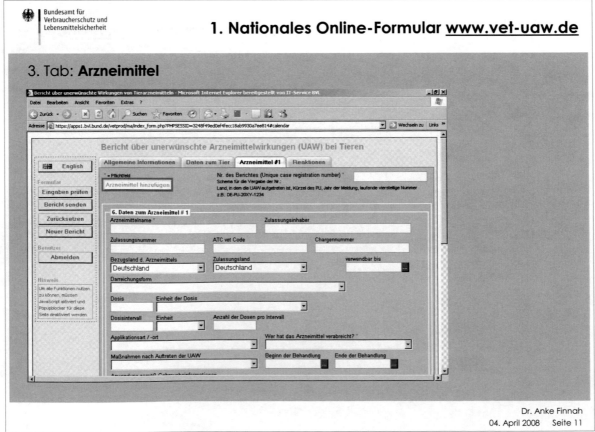

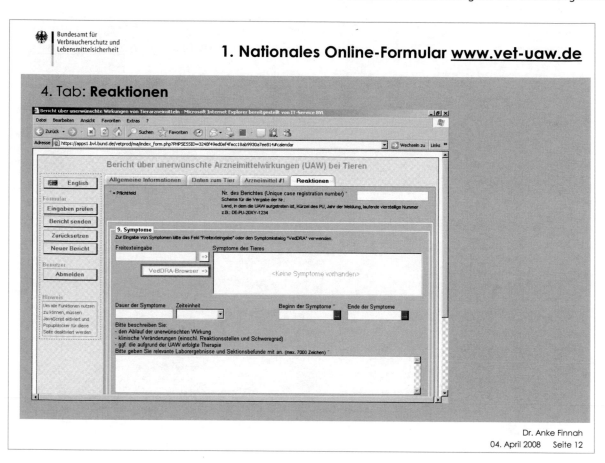

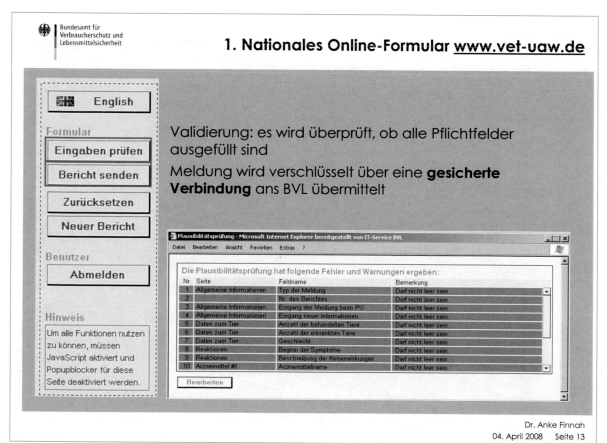

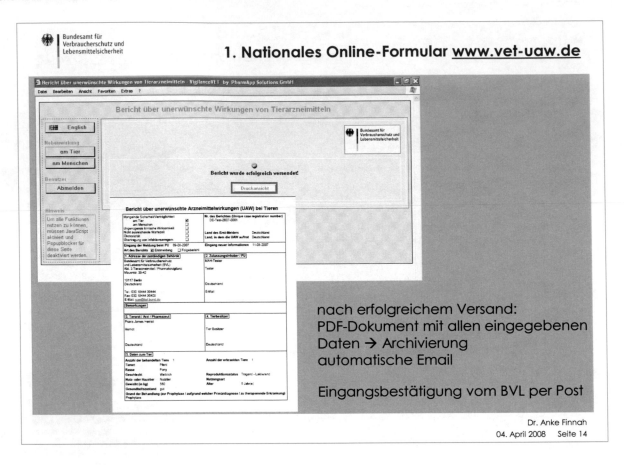

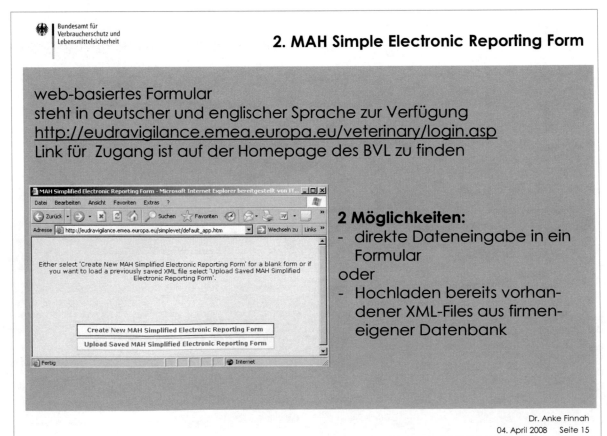

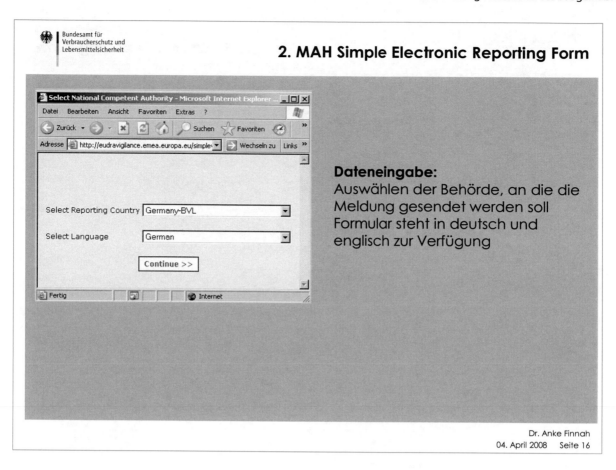

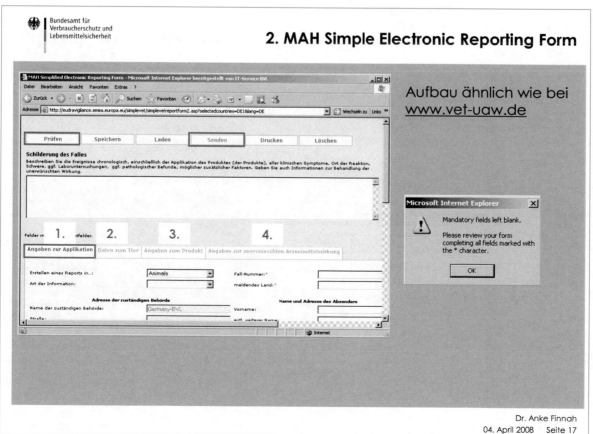

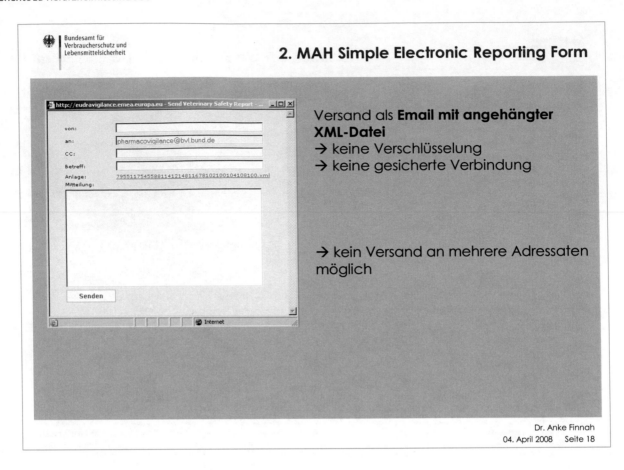

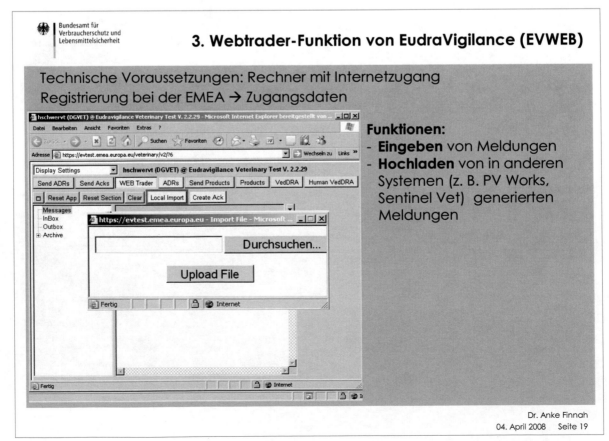

Bundesamt für
Verbraucherschutz und
Lebensmittelsicherheit

3. Webtrader-Funktion in EudraVigilance (EVWEB)

Auswahl der Adressaten aus einer Liste → gleichzeitiger **Versand an mehrere Adressaten** in einem Schritt möglich

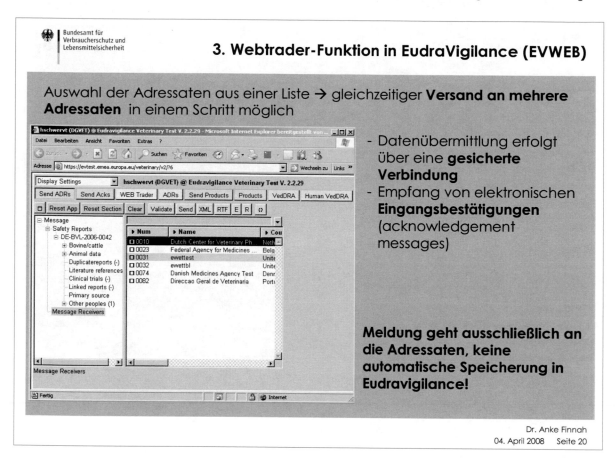

- Datenübermittlung erfolgt über eine **gesicherte Verbindung**
- Empfang von elektronischen **Eingangsbestätigungen** (acknowledgement messages)

Meldung geht ausschließlich an die Adressaten, keine automatische Speicherung in Eudravigilance!

Dr. Anke Finnah
04. April 2008   Seite 20

Bundesamt für
Verbraucherschutz und
Lebensmittelsicherheit

4. ESTRI-Gateway

in erster Linie für große, international agierende Firmen interessant
Prinzip: firmeneigene Datenbank, Versand der Informationen in Form von XML-Dateien, gleichzeitiger Versand an mehrere Adressaten möglich

Informationen und Hilfe unter
http://eudravigilance.emea.europa.eu/veterinary/evGateway01.asp
ICH M2 Gateway Recommendation for the Electronic Transfer of Regulatory Information

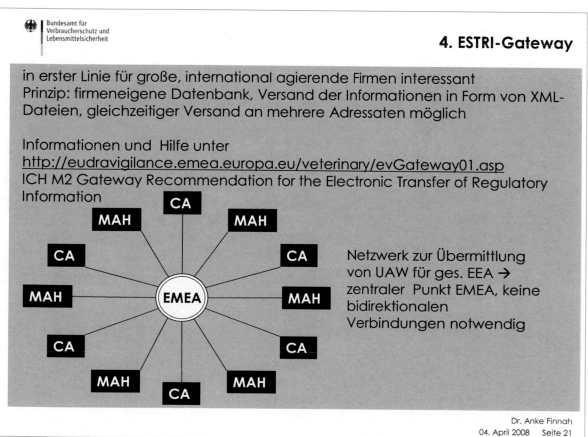

Netzwerk zur Übermittlung von UAW für ges. EEA → zentraler Punkt EMEA, keine bidirektionalen Verbindungen notwendig

Dr. Anke Finnah
04. April 2008   Seite 21

Bundesamt für
Verbraucherschutz und
Lebensmittelsicherheit

4. ESTRI-Gateway

technische Anforderungen:

- S/MIME compatible email system using POP/SMTP, direct connection via HTTP(s) or FTP
- Support for digitally signed MDN
- X.509 digital certificate support
- EDIINT/AS1 compliance certification or AS2 interoperability
- Direct transmittal of XML documents

vor dem Start Testphase mit der EMEA → Testprotokoll, dann Registrierung für Produktivsystem
Test mit dem BVL jederzeit möglich

Dr. Anke Finnah
04. April 2008   Seite 22

Bundesamt für
Verbraucherschutz und
Lebensmittelsicherheit

Ausblick

Workshop/Erfahrungsaustausch:
- Elektronisches Melden
- Causality-Assessment
- VedDRA-Codierung
- EVWEB/Webtrader (wenn genügend Teilnehmer evtl. alternativ zu Termin bei der EMEA)
- nationales UAW-Meldeformular

im BVL, Berlin
Zeitraum: a.s.a.p.

Themenvorschläge?

Kontakt:
uaw@bvl.bund.de
030/18444-30444

Dr. Anke Finnah
04. April 2008   Seite 23

6 Regulatory Aspects of Marketing Authorization of Ectoparasiticides for Dogs and Cats in Germany

Stephan Steuber

Federal Office of Consumer Protection and Food Safety, Berlin

Correspondence to: Dr. S. Steuber, BVL, Ref. 303, Mauerstraße 39–42, D-10117 Berlin, Germany, Tel. 030 18 444 30313, e-mail: Stephan.Steuber@bvl.bund.de

Bundesamt für Verbraucherschutz und Lebensmittelsicherheit

Regulatory Aspects of Marketing Authorization of Ectoparasiticides for Dogs and Cats

1. Current options of licensing
2. General requirements on efficacy
3. Specific EU-Guidelines for ectoparasiticidal products in cats and dogs
4. Supportive NGO-Guidelines (technical guidance published by WAAVP)
5. EU-Guideline on User Safety (topical products)
6. Summary

Stephan Steuber • 19. November 2007 • 2

Bundesamt für
Verbraucherschutz und
Lebensmittelsicherheit

1. Current options of licensing

Authorization procedures for ectoparasitic products

✓   **Licensing as a** Biocidal Product (BP)
*according to the German Chemical Act and
the rules laid down in Article 5 of Directive 98/8/EC.*

✓   **Marketing Authorization as a**
Veterinary Medicinal Products (VMP)
*at national level according to the German Medicines Act
(section 21)*

or
*at European level via a centralised (CP), decentralised (DCP) or
mutual recognition procedure (MRP).*

Stephan Steuber • 19. November 2007 • 3

Bundesamt für
Verbraucherschutz und
Lebensmittelsicherheit

1. Current options of licensing

Definition: BP according to Directive 98/8/EC

- Active substances and preparations containing one or more active substances, put up in the form in which they are supplied to the user, intended to destroy, deter, render harmless, prevent the action of, or otherwise exert a controlling effect on any harmful organism by chemical or biological means.

Definition: VMP according to Directive 2001/82/EC

- Any substance or combination of substances presented for treating or preventing disease in animals. Any substance or combination of substances which may be administered with a view to making a medical diagnosis or to restoring, correcting or modifying physiological functions in animals.

Stephan Steuber • 19. November 2007 • 4

Bundesamt für
Verbraucherschutz und
Lebensmittelsicherheit

1. Current options of licensing

Biocidal Products (BPs):

- Products having only repellent activity, without any killing effect (e.g. collars, neckties, ear marks), or repellents (without a killing effect and a therapeutic claim) that are directly applied to animal skin, are considered biocidal products.

Veterinary Medicinal Products (VMPs):

- Products with <u>lethal effects</u> on external parasites to be used on animals and claiming a <u>precise therapeutic indications</u> (including prevention or treatment) are at present considered and authorised as veterinary medicinal products.

  Note: In case of uncertainty the applicant should consult the competent authorities (BVL, BAUA) for borderline setting between BP and VMP.

Stephan Steuber • 19. November 2007 • 5

Bundesamt für
Verbraucherschutz und
Lebensmittelsicherheit

1. Current options of licensing

1.1 Marketing authorization as a Biocidal Product

Responsible Authorization Office:
Federal Institute for Occupational
Safety and Health, Dortmund

The conditions for authorization are laid down in Art. 5 of Directive 98/8/EC

| 24.4.98 | EN | Official Journal of the European Communities | L 123/1 |

DIRECTIVE 98/8/EC OF THE EUROPEAN PARLIAMENT AND OF THE COUNCIL

of 16 February 1998

concerning the placing of biocidal products on the market

Stephan Steuber • 19. November 2007 • 6

Bundesamt für
Verbraucherschutz und
Lebensmittelsicherheit

1. Current options of licensing

1.2 Marketing authorization as a Veterinary Medicinal Product (VMP)

Responsible national authorization office:
Federal Office of Consumer
Protection and Food Safety (BVL), Berlin

Bundesamt für
Verbraucherschutz und
Lebensmittelsicherheit

Responsible for the scientific evaluation
of applications for European marketing
authorization: European Medicines Agency (EMEA), London

European Medicines Agency

emea

The conditions for authorization are laid down nationally in the German Drug Act and at European level in the Directive 2001/82/EC as amended (2004/28/EC).

28.11.2001   EN   Official Journal of the European Communities   L 311/1

DIRECTIVE 2001/82/EC OF THE EUROPEAN PARLIAMENT AND OF THE COUNCIL

of 6 November 2001

on the Community code relating to veterinary medicinal products

Stephan Steuber • 19. November 2007 • 7

Bundesamt für
Verbraucherschutz und
Lebensmittelsicherheit

2. General requirements on efficacy

Directive 2001/82: Annex 1

L 311/28   EN   Official Journal of the European Communities   28.11.2001

ANNEX I

REQUIREMENTS AND ANALYTICAL PROTOCOL, SAFETY TESTS, PRE-CLINICAL AND CLINICAL FOR TESTS OF VETERINARY MEDICINAL PRODUCTS

In assembling the dossier for application for marketing authorization, applicants shall take into account the Community guidelines relating to the quality, safety and efficacy of veterinary medicinal products published by the Commission in *The rules governing medicinal products in the European Community.*

All information which is relevant to the evaluation of the medicinal product concerned shall be included in the application, whether favourable or unfavourable to the product. In particular, all relevant details shall be given of any incomplete or abandoned test or trial relating to the veterinary medicinal product. Moreover, after marketing authorization, any information not in the original application, pertinent to the benefit/risk assessment, shall be submitted forthwith to the competent authority.

Stephan Steuber • 19. November 2007 • 8

Bundesamt für
Verbraucherschutz und
Lebensmittelsicherheit

2.1 General requirements on efficacy: Annex 1 Directive 2001/82

PART 4

Pre-clinical and clinical testing

The particulars and documents which shall accompany applications for marketing authorizations pursuant to Articles 12(3)(j) and 13(1) shall be submitted in accordance with the provisions of this Part.

A.   PHARMACOLOGY

A.1. *Pharmacodynamics*

The study of pharmacodynamics shall follow two distinct lines of approach:

First, the mechanism of action and the pharmacological effects on which the recommended application in practice is based shall be adequately described. The results shall be expressed in quantitative terms (using, for example, dose-effect curves, time-effect curves, etc.) and, wherever possible, in comparison with a substance the activity of which is well known. Where a higher efficacy is being claimed for an active substance, the difference shall be demonstrated and shown to be statistically significant.

A.2. *Pharmacokinetics*

Basic pharmacokinetic information concerning a new active substance is generally useful in the clinical context.

Bundesamt für
Verbraucherschutz und
Lebensmittelsicherheit

2.1 General requirements on efficacy: Annex 1 Directive 2001/82

PART 4

Pre-clinical and clinical testing

The particulars and documents which shall accompany applications for marketing authorizations pursuant to Articles 12(3)(j) and 13(1) shall be submitted in accordance with the provisions of this Part.

B. TOLERANCE IN THE TARGET SPECIES OF ANIMAL

The purpose of this study, which shall be carried out with all animal species for which the medicinal product is intended, is to carry out in all such animal species local and general tolerance trials designed to establish a tolerated dosage wide enough to allow an adequate safety margin and the clinical symptoms of intolerance using the recommended route or routes, in so far as this may be achieved by increasing the therapeutic dose and/or the duration of treatment. The report on the trials shall contain as many details as possible of the expected pharmacological effects and the adverse reactions; the latter shall be assessed with due regard to the fact that the animals used may be of very high value.

The medicinal product shall be administered at least via the recommended route of administration.

C. RESISTANCE

Data on the emergence of resistant organisms are necessary in the case of medicinal products used for the prevention or treatment of infectious diseases or parasitic infestations in animals.

Bundesamt für
Verbraucherschutz und
Lebensmittelsicherheit

2.1 General requirements on efficacy: Annex 1 Directive 2001/82

Chapter II

Clinical requirements

1. General principles

The purposes of clinical trials are to demonstrate or substantiate the effect of the veterinary medicinal product after administration of the recommended dosage, to specify its indications and contra-indications according to species, age, breed and sex, its directions for use, any adverse reactions which it may have and its safety and tolerance under normal conditions of use.

Unless justified, clinical trials shall be carried out with control animals (controlled clinical trials). The effect obtained should be compared with a placebo or with absence of treatment and/or with the effect of an authorized medicinal product known to be of therapeutic value. All the results obtained, whether positive or negative, shall be reported.

The methods used to make the diagnosis shall be specified. The results shall be set out by making use of quantitative or conventional clinical criteria. Adequate statistical methods shall be used and justified.

Experimental data shall be confirmed by data obtained under practical field conditions.

Stephan Steuber • 19. November 2007 • 11

Bundesamt für
Verbraucherschutz und
Lebensmittelsicherheit

2.1 General requirements on efficacy: Annex 1 Directive 2001/82

2. Performance of trials

All veterinary clinical trials shall be conducted in accordance with a fully considered detailed trial protocol which shall be recorded in writing prior to commencement of the trial. The welfare of the trial animals shall be subject to veterinary supervision and shall be taken fully into consideration during the elaboration of any trial protocol and throughout the conduct of the trial.

In the case of fixed combination products, the investigator shall also draw conclusions concerning the safety and the efficacy of the product when compared with the separate administration of the active substances involved.

3. Concluding expert report

The concluding expert report shall provide a detailed critical analysis of all the pre-clinical and clinical documentation in the light of the state of scientific knowledge at the time the application is submitted together with a detailed summary of the results of the tests and trials submitted and precise bibliographic references.

Stephan Steuber • 19. November 2007 • 12

3. Specific EU- Guidelines for ectoparasiticidal products

EMEA: Introductory note on Scientific Guidelines

The EMEA's Committee for Medicinal Products for Veterinary Use (CVMP) prepares Scientific Guidelines, in consultation with the competent authorities of the EU Member States, to help applicants prepare marketing-authorisation applications for veterinary medicinal products.

Guidelines are intended to provide a basis for practical harmonisation of the manner in which the EU Member States and the EMEA interpret and apply the detailed requirements for the demonstration of quality, safety and efficacy contained in the Community directives. They also help to ensure that applications for marketing authorisation are prepared in a manner that will be recognised as valid by the EMEA.

Stephan Steuber • 19. November 2007 • 13

3.1 Specific EU- Guideline: Demonstration of Efficacy of Ectoparasiticides

THE RULES GOVERNING MEDICINAL PRODUCTS IN THE EUROPEAN UNION

Volume 7: Guidelines
        Veterinary Medicinal Products

Guideline Title: Demonstration of Efficacy of Ectoparasiticides (1994), 215 -224

http://ec.europa.eu/enterprise/pharmaceuticals/eudralex/vol-7

Stephan Steuber • 19. November 2007 • 14

3.1 Specific EU- Guideline: Demonstration of Efficacy of Ectoparasiticides

1. Claiming requirements in Companion Animals:

➤ At the end of the time period as indicated by the applicant, the overall efficacy of ectoparasiticides in treating infections should be achieved as follows:

- ✓ for fleas: approximately 100%
- ✓ for lice: approximately 100%
- ✓ for mites: approximately 100% for *Sarcoptes scabiei* var. *canis* and, if possible, more than 90% for other mange mites.
- ✓ for ticks: more than 90%

☞Note! Where efficacy is less than the above no claim should be made unless the applicant can demonstrate that the degree of efficacy achieved is better than or comparable with current alternatives.

Stephan Steuber • 19. November 2007 • 15

3.1 Specific EU- Guideline: Demonstration of Efficacy of Ectoparasiticides

2. GENERAL REQUIREMENTS:

2.1 Mode of action
The pharmacodynamics of the active ingredient(s) on the target ectoparasite(s) should be adequately described.

2.2 Titration of dose
Ideally, four groups, each consisting of a sufficient number of animals to allow statistical analysis, should be administered 0, 0.5, 1 and 2 times the anticipated recommended dose. Each group should harbour or be uniformly infested with adequate numbers of each species of ectoparasites. Ideally the final formulation should be used in these trials.

2.3 Dose confirmation trials
- ✓ At least two controlled tests are recommended to demonstrate the efficacy of a new product against each ectoparasite species and stage of development as indicated in the labelling.
- ✓ Where applicable, trials should be performed in different geographic and climatic regions.
- ✓ Where applicable, at least one trial should be performed using naturally infested animals.

Stephan Steuber • 19. November 2007 • 16

3.1 Specific EU- Guideline: Demonstration of Efficacy of Ectoparasiticides

2.4 Clinical field trials

✓ Clinical field trials are required primarily for follow-up evaluation of the performance of the product as employed by the user in the field and to gain experience on the efficacy and safety of the product when applied under various clinical conditions.

✓ Field trials should be conducted in at least 2 different geographic and climatic regions.

3. SPECIAL REQUIREMENTS (Topical use Products)

✓ the effect of (artificial) rainfall at various intervals

✓ the effect of sunshine and hot weather

✓ the effect of washing and bathing

✓ the effects of hair length

✓ the effect of self-grooming

Stephan Steuber • 19. November 2007 • 17

3.2 Specific EU- guideline for testing the efficacy of antiparasitics for tick and flea infestation in dogs and cats

European Medicines Agency
Veterinary Medicines and inspections

London, 12 November 2007
EMEA/CVMP/EWP/005/2000-Rev.2

**COMMITTEE FOR MEDICINAL PRODUCTS FOR VETERINARY USE
(CVMP)**

**GUIDELINE FOR THE TESTING AND EVALUATION OF THE EFFICACY OF
ANTIPARASITIC SUBSTANCES FOR THE TREATMENT AND PREVENTION OF
TICK AND FLEA INFESTATION IN DOGS AND CATS**

1.   SCOPE

This note provides special guidance with respect to the testing and evaluation of efficacy of veterinary antiparasitic products that are intended for external use for the treatment and prevention of tick and flea infestations, or systemic use for the treatment and prevention of flea infestations in dogs and cats. Information is also provided for the testing of veterinary antiparasitic products containing substances with insect growth regulating properties IGRs), either as mono-preparations or in combination with a flea adulticide.

Stephan Steuber • 19. November 2007 • 18

Bundesamt für
Verbraucherschutz und
Lebensmittelsicherheit

3.2 Specific EU- guideline for testing the efficacy of antiparasitics for tick and flea infestation in dogs and cats

3.1 Ectoparasite species

Most relevant tick and flea species in dogs and cats in Europe:

Ticks:

Dermacentor reticulatus
Ixodes hexagonus
Ixodes ricinus
Rhipicephalus sanguineus (dogs)

Fleas:
Ctenocephalides canis (dogs)
Ctenocephalides felis

4. STUDY DESIGN FOR TESTING THE EFFICACY OF PRODUCTS FOR THE TREATMENT AND PREVENTION OF TICK INFESTATION

Studies for each tick species and each stage of the life cycle against which efficacy is claimed should be provided. The applicant should justify the type of studies (*in vitro* and *in vivo* laboratory studies and field studies) for each species and stage.

Bundesamt für
Verbraucherschutz und
Lebensmittelsicherheit

3.2 Specific EU- guideline for testing the efficacy of antiparasitics for tick and flea infestation in dogs and cats

4.1.5 Criteria of efficacy

4.1.5.1 Repellent effect

A repellent effect means that no ticks will attach to the animal. Ticks already on the animal will leave the animal soon after treatment.
In general, no ticks should be detectable on the animal after 24 hours following administration of the product.

4.1.5.2 Acaricidal effect

In evaluating the acaricidal efficacy of a product, the feeding or engorgement of ticks should be taken into consideration in addition to the rate at which ticks are killed. It is recommended to assess the acaricidal effect on individual ticks according to the following parameters:

Bundesamt für
Verbraucherschutz und
Lebensmittelsicherheit

3.2 Specific EU- guideline for testing the efficacy of antiparasitics for tick and flea infestation in dogs and cats

Definition of parameters

Category	General findings	Attachment status	Effect
1	Live	Free	No
2	Live	Attached; unengorged	No (except single ticks)
3	Live	Attached; engorged	No (except single ticks) *
4	Killed	Free	Yes
5	Killed	Attached; unengorged	Yes
6	Killed	Attached; engorged	No (except single ticks) *

Engorgement of ticks is indicated by the presence of blood in the digestive tract. Engorgement can be determined visually by squeezing technique or microscopically or by accurate weighing

Suggested SPC warnings:

There may be an attachment of single ticks. For this reason, a transmission of infectious disease by ticks cannot be completely excluded if conditions are unfavourable.

Ticks will be killed and fall off the host within 24 to 48 hours after infestation without having had a blood meal, as a rule. An attachment of single ticks after treatment cannot be excluded.

 Bundesamt für
Verbraucherschutz und
Lebensmittelsicherheit

3.2 Specific EU- guideline for testing the efficacy of antiparasitics for tick and flea infestation in dogs and cats

5. STUDY DESIGN FOR TESTING THE EFFICACY OF PRODUCTS FOR THE TREATMENT AND PREVENTION OF FLEA INFESTATION

Both laboratory and field studies should be performed for each animal species claimed (dog/cat).

The efficacy of the proposed product should be <u>at least 95% for adult fleas</u> at each counting during the claimed efficacy period.

5.1.6 Testing for photostability

For products intended for external use, the final formulation intended for marketing should be tested for its photostability, e.g. by UV radiation according to <u>VICH GL5</u> (photostability testing of new

5.1.7 Testing for water stability

For products intended for external use, the water stability of the formulation intended for marketing should be demonstrated, especially for products with a claimed duration of efficacy for 2 weeks or more. <u>The impact of exposure to water e.g. through shampooing, swimming or rainwater on the insecticidal effect should be evaluated at regular intervals (e.g. once a week)</u>. Alternatively, data on the concentration time course of the active substance in the fur after single/repeated washing can be provided.

Bundesamt für
Verbraucherschutz und
Lebensmittelsicherheit

3.2 Specific EU- guideline for testing the efficacy of antiparasitics for tick and flea infestation in dogs and cats

New: From 2008 onwards

5.3 Specific recommendations for efficacy testing of veterinary medicinal products containing insect-growth regulators (IGRs) against fleas

The use of IGRs in cats or dogs is limited to the treatment and prevention of flea infestations. Although it is acknowledged that some few IGRs could also affect ticks, IGRs are not considered suitable in the treatment and prevention of tick infestations, because the tick species common in Europe (*Dermacentor reticulatus, Ixodes ricinus, I. hexagonus, Rhipicephalus sanguineus*) are three-host ticks. Laboratory and field studies demonstrating the IGR properties should be provided. The

5.3.1.1.1 *In vitro* studies to demonstrate ovicidal activity

The effect of an insect growth regulator *via* contact on flea metamorphosis (sterilisation of eggs/ inhibition of egg hatching and the formation of cocoons) should be demonstrated and the LC_{50} and LC_{90} calculated, using justified recognized methods.

5.3.1.1.2 *In vitro* studies to demonstrate larvicidal activity

To determine the larvicidal LC_{50} and LC_{90} of an insect growth regulator *in vitro* (e.g. juvenile hormone antagonist) preferably 2nd or early 3rd instar larvae of a well established flea strain should be used because of convenience in handling. In order to adequately calculate dose effect relationship of an IGR, it is recommended to use at least 20 -50 viable larvae at each test concentration.

Stephan Steuber • 19. November 2007 • 23

Bundesamt für
Verbraucherschutz und
Lebensmittelsicherheit

3.2 Specific EU- guideline for testing the efficacy of antiparasitics for tick and flea infestation in dogs and cats

5.3.1.2 *In vivo* studies

In case of a combination product containing both an IGR and an adulticidal, the demonstration of the IGR efficacy may be markedly impeded by the rapid killing effect of the adulticidal compound. In such a case it may be necessary to increase the number of fleas for infestations in the controlled study according to the WAAVP guidelines (e.g. 200/animal) and/ or extend the study period in order to generate adequate numbers of eggs for the calculation of the ovicidal activity. Reinfestations should

The efficacy of the proposed product should be at least 95% for adult fleas and at least 90 % for the inhibition of the emergence to adults (IGR).

5.3.2 Specific Field trial recommendations for Insect Growth Regulator (IGR)

If prevention of flea multiplication by inhibiting egg development is claimed only (e.g. an IGR mono product), the study should be performed on animals harbouring apparently no or low numbers of fleas (0 – 3 fleas/animals) at the commencement of the trial period. During the study any concomitant

If treatment and prevention of flea infestations (e.g. product combining an IGR and an adulticidal) is claimed, animals enrolled in the study should harbour a natural flea burden of at least 5 to 10 fleas per animal on average. Appropriate control should be included, e.g. an approved adulticidal product alone or a fixed combination product of an adulticidal and an insect growth regulator.

Stephan Steuber • 19. November 2007 • 24

Bundesamt für
Verbraucherschutz und
Lebensmittelsicherheit

4. Supportive NGO-Guidelines (WAAVP, 2007)

Available online at www.sciencedirect.com

ScienceDirect

Veterinary Parasitology 145 (2007) 332–344

veterinary parasitology

www.elsevier.com/locate/vetpar

World Association for the Advancement of Veterinary Parasitology
(W.A.A.V.P.) guidelines for evaluating the efficacy of parasiticides
for the treatment, prevention and control of flea and tick
infestation on dogs and cats

A.A. Marchiondo [a,*], P.A. Holdsworth [b], P. Green [c], B.L. Blagburn [d], D.E. Jacobs [e]

„These guidelines is intended to assist planning and conduct of laboratory and clinical studies to assess the efficacy of ectoparasiticidal applied to dogs and cats.
They are also intended to assist registration authorities involved in the approval and registration of new parasiticides and to facilitate the worldwide adoption of harmonized procedures."

Stephan Steuber • 19. November 2007 • 25

Bundesamt für
Verbraucherschutz und
Lebensmittelsicherheit

4. Supportive NGO-Guidelines (WAAVP, 2007)

Supranational approach:

3. Fleas and ticks on dogs and cats

Table 1
Main flea species on cats and/or dogs

Flea species	Australia	Europe	Japan	USA	Southern Africa
Ctenocephalides felis	X	X	X	X	X
Ctenocephalides canis	–	X	X	–	–

334   A.A. Marchiondo et al. / Veterinary Parasitology 145 (2007) 332–344

Table 2
The main ixodid tick species that feed on dogs and cats

Tick species	Australia	Europe	Japan	USA	Southern Africa
Amblyomma americanum	–	–	–	X	–
Amblyomma triguttatum	X	–	–	–	–
Dermacentor variabilis	–	–	–	X	–
Dermacentor reticulatus	–	X	–	–	–
Haemaphysalis leachi					X
Haemaphysalis longicornis	Rare	–	X	–	–
Haemaphysalis flava	–	–	X	–	–
Hyalomma bursa	–	X	–	–	–
Ixodes hexagonus	–	X	–	–	–
Ixodes holocyclus	X	–	–	–	–
Ixodes ovatus	–	–	X	–	–
Ixodes ricinus	–	X	–	–	–
Ixodes scapularis	–	–	–	X	–
Rhipicephalus sanguineus	X	X	X	X	–
Rhipicephalus bursa	–	X	–	–	–

Stephan Steuber • 19. November 2007 • 26

Bundesamt für
Verbraucherschutz und
Lebensmittelsicherheit

5. EU-Guideline on user safety (topical products)

GUIDELINE ON USER SAFETY

FOR PHARMACEUTICAL VETERINARY MEDICINAL PRODUCTS

2   Scope

This guideline applies to all new applications for Marketing Authorisation for pharmaceutical veterinary medicinal products. This guideline shall not apply to Marketing Authorisations granted in

The assessment of the user safety will comprise the following steps:

Exposure assessment → hazard identification and characterisation → risk characterisation → risk management

5   Hazard identification and characterisation

- Both local and systemic effects should be considered. Whenever possible, dose-response relationships have to be identified in order to derive the no observed adverse effect level (NOAEL), or, if this is not possible, the lowest observed adverse effect level (LOAEL).

- Toxicity studies should employ the same routes of exposure as described in the exposure scenarios. Alternatively, route-to-route extrapolation may be considered when appropriate. It

Stephan Steuber • 19. November 2007 • 27

Bundesamt für
Verbraucherschutz und
Lebensmittelsicherheit

5. EU-Guideline on user safety (topical products)

6   Risk characterisation

6.2   Quantitative risk characterisation

6.2   Quantitative risk characterisation

The procedure for the quantitative risk assessment consists of comparing the exposure levels to which the user is exposed or is likely to be exposed with the exposure levels at which no adverse effects are expected to occur. This is generally done by comparing the estimated exposure to the relevant NOAEL.

Where the exposure estimate is higher than or equal to the NOAEL, the risk for the user is considered to be unacceptably high.

Where the exposure estimate is less than the NOAEL, the magnitude by which the NOAEL exceeds the estimated exposure (i.e. the margin of exposure (MOE)) needs to be considered taking account of the following parameters:
- the intra- and interspecies variation;

- the nature and severity of effect;

- the human population to which the exposure information applies;

- the differences in exposure (route, duration, frequency);

- the dose-response relationship observed;

Stephan Steuber • 19. November 2007 • 28

Bundesamt für
Verbraucherschutz und
Lebensmittelsicherheit

5. EU-Guideline on user safety (topical products)

7.3 Risk control options

In general, the following options for risk control may be used:

- restriction of the distribution, e.g. as prescription only medicine;

- excluding groups at risk, e.g. sensitised people, pregnant women;

- restriction of application methods, e.g. pour-on instead of spraying or use of closed delivery systems;

- restriction of the field of use, e.g. outdoor use only;

- modification of the formulation, e.g. ready-to-use rather than concentrate, replacement of substances of concern with less dangerous ones, etc.;

- modification of packaging, e.g. reduced pack size;

8   Risk communication

Example:

- This product can cause eye-irritation (A).

- Avoid contact with the eyes (B).

- Wear protective glasses (C).

- When the product comes into contact with the eyes, rinse immediately with plenty of water (D).

Bundesamt für
Verbraucherschutz und
Lebensmittelsicherheit

5. EU-Guideline on user safety (topical products)

SPC warnings Spot-on's
Example: ProMeris Duo™ Spot-on
(Actives: Metaflumizone/Amitraz)

➢ "This product contains amitraz, which can have harmful effects in humans. Children should not have access to used pipettes. Used pipettes should be disposed of immediately."

➢ "Avoid direct contact with treated animals until the application site is dry. Children should not be allowed to play with treated animals until the application site is dry. It is therefore recommended to treat the animals during the evening and that recently treated animals are not allowed to sleep with owners, especially children."

6. Summary

1. **Products containing active substances with lethal effects on external parasites to be used on animals are in general considered and authorized as VMP with precise veterinary medicinal indications (including prevention, treatment or diagnosis of a disease).**

2. **On a national level the competent authority for grant marketing authorizations is the Federal Office of Consumer Protection and Food Safety (BVL). The conditions for authorization are laid down in the German Drug Act having regard to the European Directive 2001/82/EC as amended (2004/28/EC).**

3. **The responsible office for the scientific evaluation of applications for European marketing authorizations (Centralised Procedures) is the European Medicines Agency (EMEA).**

Stephan Steuber • 19. November 2007 • 31

6. Summary

5. **Specific guidelines for ectoparasiticidal products in companion animals have been generated to provide a practical basis for execution of studies requested for marketing authorization.**

6. **Guidelines do also provide guidance for member states within the EU to harmonise the requirements for demonstration efficacy in companion animals.**

7. **For topical ectoparasiticidal products the EU - User Safety Guideline should also be considered.**

8. **Supranationally (VICH-level), guidelines for ectoparasiticidal products in companion animals are still pending (in contrast to guidelines for anthelmintics).**

Stephan Steuber • 19. November 2007 • 32